The Adventures of Ellie and Ollie

Tree Treasure

Marina Swan

Ellie and Ollie were playing in the backyard when something sparkling caught Ellie's eye. "Ollie! Come and look!"

Ellie stared up at the tree.
Tiny flashes of green, blue, orange
and gold shimmered between the
leaves.

Her eyes grew wide.
"Woah! That tree has TREASURE!" she cried.

Ollie came over and looked closely.
He smiled. "It does look like treasure,
Ellie. But, those are jewels."

Ellie blinked. “JEWELS on a tree?
That’s a lucky tree.”
Ollie laughed. “Well... jewel bugs!”

"Wow, jewel bugs! They are so colourful and shiny! No wonder they're called jewel bugs!"

"Their beautiful colours aren't just for looking pretty," Ollie explained. "They warn other animals to leave them alone."

"What about the gold ones?" Asked Ellie. "Well the gold ones are the babies." "Wow golden babies!"

"Baby bugs are called nymphs, and their colours can change as they grow." Said Ollie. Ellie crept closer to get a better look.

And closer. And just a little closer...
Suddenly, she wrinkled her nose.
"Ewww! What's so STINKY?"

"That is the jewel bugs. They can make a smell when they feel frightened." Ellie's ears drooped. "Did I scare them?" "Maybe you got a little too close," Ollie said gently.

Ellie quickly stepped backwards. "Sorry, little jewel bugs. I didn't mean to scare you." From a little farther away, she could still see their beautiful colours sparkling in the sunshine.

Ellie and Ollie wandered over to another tree. Ellie searched the leaves. "There are only a few jewel bugs on this one!" They checked another. Only a couple!

Ellie looked back at the first tree. "But there are HEAPS on that one! Why?" "Jewel bugs gather on plants that are healthy and they can find all the food and shelter they need," Ollie explained.

They wandered back to the jewel bug tree. This time, Ellie didn't rush closer. She sat beside Ollie and watched quietly. The little bugs crawled over the leaves and branches, sparkling in the sunlight.

"They really are like tiny jewels," Ellie whispered. Then she smiled. "But I don't need to pick up treasure to know it's special."

Ollie looked at her proudly.
"And some treasure is best left right where you find it."

About the Author

Marina Swan has spent over 20 years as an early years teacher, sharing her love of stories with children. She longed for books that connected more closely to what kids were learning in school—so she began writing her own.

Her stories bring the early years curriculum to life, weaving reading, writing, math, science, history, and social skills into joyful adventures. Designed for both home and classroom, they blend learning with imagination.

Inspired by her ginger cat and a close friend's golden retriever, Marina's characters are real and relatable, helping children see themselves in every story. Her passion is simple: to spark a love of learning through stories children will treasure.

If you would like to feature on our pages, send in your photos and videos while you are exploring Ellie and Ollie's Adventures!

A Special Thank You

A very special thank you to Prep D, whose kindness and care for the little creatures around them helped inspire this story.

The children noticed that some jewel bugs were being taken from the trees and placed into containers and bottles. Instead of simply walking past, they wondered what they could do to help. They decided that Ellie and Ollie could share an important message about enjoying nature while leaving living things safely where they belong.

Ellie and Ollie loved working with Prep D to create this story! Thank you for your wonderful ideas, curiosity and, most of all, your thoughtfulness.

Ellie and Ollie hope their adventure will help everyone at school remember that we can look, wonder and learn about the amazing creatures around us – while respecting them and leaving them where they belong.

Discussion & Activity Questions

Why do you think Ellie thought the tree was full of treasure when she first saw the jewel bugs? What colours would you expect to see on a jewel bug?

Ollie taught Ellie that jewel bugs can have bright, shiny colours. Why do you think their colours might help protect them from other animals?

Ellie got a little too close and noticed a stinky smell! Why do jewel bugs make a smell when they feel threatened? What do you think Ellie should do when she notices the bugs are scared?

Ellie and Ollie noticed lots of jewel bugs on one tree and only a few on the other trees. Why do you think jewel bugs might choose some plants or trees over others?

Ellie learned that she could enjoy the jewel bugs without picking them up. What are some ways we can look, wonder and learn about nature while still leaving little creatures where they belong?

If you could ask Ellie or Ollie a question about jewel bugs, what would it be?

Other books in this series:

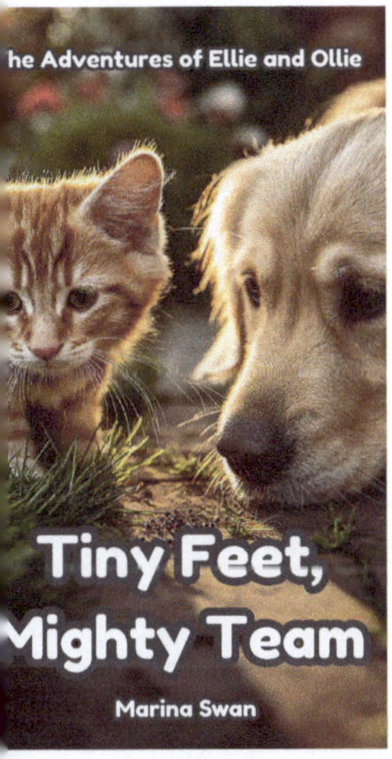

Other books in this series:

www.ingramcontent.com/pod-product-compliance
Lightning Source LLC
LaVergne TN
LVHW070626061126
851857LV00002B/19